Starship's Fifth Flight Test

Catching Boosters and Landing Dreams

A Gripping Look at SpaceX's Bold Journey into Rocket Reusability and Atmospheric Mastery

Joe E. Grayson

Copyright © 2024 Joe E. Grayson, All rights reserved.

No part of this publication may be reproduced, distributed, or transmitted in any form or by any means, including photocopying, recording, or other electronic or mechanical methods, without the prior written permission of the publisher, except in the case of brief quotations embodied in critical reviews and certain other noncommercial uses permitted by copyright law.

Table of Contents

Table of Contents
Introduction
Chapter 1: Countdown to Launch
Chapter 2: Liftoff and Initial Ascent
Chapter 3: Booster Separation and Boost-Back Burn
Chapter 4: The High-Stakes Catch Decision
Chapter 5: The Booster Return and Catch Attempt
Chapter 6: Starship's Journey Beyond Booster Separation
Chapter 7: Atmospheric Re-Entry and Heat Management
Chapter 8: Final Descent and Targeted Landing
Chapter 9: Engineering Triumphs and Lessons Learned
Chapter 10: The Road Ahead for SpaceX and Starship
Conclusion

Introduction

SpaceX's journey toward reusability has been nothing short of revolutionary, reshaping the future of space travel by making the extraordinary goal of returning rockets back to Earth not just possible but practical. Since the early days of rocketry, sending objects into space has been one of humanity's greatest achievements—and greatest expenses. Rockets, designed for single-use launches, would be discarded after delivering their payload, a process as costly as it was wasteful. SpaceX, however, emerged with a vision that challenged this tradition, determined to create a fully reusable launch system. This aspiration wasn't merely about saving costs; it was about changing the entire paradigm of space exploration, making frequent, affordable space travel a reality.

Central to SpaceX's mission is the Starship program, a bold endeavor to develop a spacecraft capable of carrying large numbers of people and heavy cargo into space, potentially to other planets. Unlike previous rockets, Starship is designed for complete reusability, from launch to landing. This innovation has the potential to open new doors in exploration, transforming space into an accessible frontier for scientific research, tourism, and even colonization. Starship isn't merely a rocket; it's a vision for the future, one that could change life on Earth by bringing us closer to becoming an interplanetary species.

The fifth flight test represents a critical chapter in SpaceX's quest for reusability. It's a mission layered with complexity and high stakes, testing not only the technology but the ingenuity and resilience of the engineers and team behind it.

This particular test aimed to demonstrate new aspects of Starship's capabilities, including the unprecedented attempt to catch the booster upon its return to Earth. It was a daring maneuver that, if successful, would mark an extraordinary milestone in reusability and precision landing.

This book delves into the intricacies of that fifth flight test, examining each stage, from the tense moments before liftoff to the breathtaking conclusion. Readers will gain insight into the advanced technologies, strategic decisions, and real-time problem-solving that made this test a monumental achievement. By exploring SpaceX's successes and challenges, this journey reveals the broader impact of the Starship program on the future of space travel and the role it may play in humanity's ultimate quest: to venture beyond our planet. This story isn't only about a single test; it's a glimpse into the daring spirit that drives us toward the stars.

Chapter 1: Countdown to Launch

In the hours leading up to launch, SpaceX's team meticulously prepared for every detail of the fifth Starship flight test, a mission that represented months of planning and countless hours of teamwork. Each stage of pre-flight preparations was a coordinated effort involving numerous systems, checks, and re-checks to ensure that every component of the rocket, from its massive Super Heavy booster to its complex array of Raptor engines, was primed for success. This wasn't just about equipment readiness; it was a dance of precision, with each team member playing an essential role in a finely tuned sequence that would culminate in liftoff.

The preparations began with a series of technical verifications on the launch vehicle itself, where engineers examined everything from fuel levels to engine functionality. This phase involved inspecting each Raptor engine—behemoth powerhouses designed to withstand the tremendous forces of launch. Technicians worked carefully through diagnostic routines, ensuring that each engine was capable of delivering the necessary thrust without issue. Fueling the rocket, an intricate operation of its own, involved careful monitoring to control temperature and pressure levels of the cryogenic fuel, critical for ensuring efficient engine performance. Each reading was checked, double-checked, and then reviewed again, a process that highlighted the unwavering commitment to safety and precision.

Meanwhile, the control team, seated in the mission control center, reviewed their launch sequences and communications. In these tense moments before liftoff, the team rehearsed scenarios repeatedly, from nominal launches to potential malfunctions, to ensure everyone was prepared for any outcome. The flight director led the team in each phase, coordinating a complex web of specialists responsible for different aspects of the mission—from propulsion and navigation to telemetry and safety. It was their job to monitor the real-time data from hundreds of sensors onboard the vehicle, instantly analyzing each variable and ensuring that all systems remained within operational limits.

Outside of mission control, the support crew prepared ground equipment and infrastructure, carefully inspecting the launch pad to ensure it could support the intense vibrations and heat that would be unleashed upon ignition. This team's role was as essential as any onboard system, as the safety of both the vehicle and ground personnel depended on their detailed preparations.

Every crew member, whether they were in mission control or on the ground, understood that even the smallest oversight could jeopardize the entire mission. With so many moving parts, each team member's expertise was critical to the mission's success. They communicated with precision, often in short, clear codes that left no room for ambiguity. Hours of pre-launch routines were practiced over and over, reflecting the extensive training and professionalism that SpaceX required for such high-stakes endeavors.

These pre-flight preparations weren't just about checking items off a list; they were the collective work of dedicated individuals, each aware of the monumental effort involved and the potential history in the making. The commitment and precision of each crew member were a testament to the discipline and unity within SpaceX, and as they reached the final go/no-go poll, the culmination of their work became apparent. Every team member had given their all to bring the Starship flight test to the brink of liftoff, each one understanding that their role was essential to the dream of reusability and the future of space travel.

As the countdown clock edged closer to zero, the intensity inside SpaceX's mission control room grew palpable. In those final moments before launch, the go/no-go poll became the crucial test of readiness—a series of coordinated affirmations from each team involved in the mission. The go/no-go poll is more than a procedural formality; it's a system of accountability and assurance, a synchronized decision-making process that underscores the importance of collective expertise in these high-stakes moments. Each team leader, responsible for a specific aspect of the launch, must confirm their area is fully operational and ready for the test. There is no room for assumptions—each voice represents a layer of safety, a final check ensuring that all elements are functioning as planned.

The poll begins with the flight director, who moves systematically through the list, calling out each system in turn: propulsion, guidance, navigation, communications, and so on. With each prompt, the respective specialist responds, "go," or, in rare cases, "no-go," signaling if there's a last-minute concern that could jeopardize the mission. These moments demand absolute clarity and focus. Team members must trust their equipment, their training, and each other, as any hesitation or miscommunication could lead to an aborted launch. It's a weight that every team member feels, knowing that their "go" will allow the mission to proceed, while a "no-go" will halt it in its tracks.

The atmosphere in mission control is one of intense concentration. Multiple screens display streams of data in real time, from fuel pressure readings and wind speeds to engine temperature metrics and trajectory models. Each specialist monitors their area closely, poised to catch any anomaly that might arise in those final seconds. Voices are clear and steady, but there's an underlying tension as each confirmation brings the team one step closer to liftoff. The flight director, at the center of it all, maintains a calm yet commanding presence, guiding the team through these critical seconds, ready to make the final call.

This tension in the room isn't just about technical details; it's a testament to the sheer dedication of everyone involved. Each specialist knows that the weight of countless hours, the collective work of the team, rests on this brief series of "go" confirmations. For some, it's years of work coming down to these last, breath-holding moments. As the final "go" is called, a ripple of relief and determination sweeps through the room. The go/no-go poll has confirmed the mission's readiness, and everyone understands that the true test has only just begun.

These moments leading up to launch define the essence of SpaceX's approach to space exploration—meticulous, disciplined, and guided by a deep respect for the power of technology and the unknowns of space. The countdown continues, and with the go/no-go poll behind them, the team braces for the roar of ignition, knowing that they are now on the edge of history.

Chapter 2: Liftoff and Initial Ascent

The launch sequence began with a surge of raw power as the Raptor engines roared to life, releasing a blast of energy that propelled the massive Starship vehicle from the pad and into the sky. In an instant, the vehicle transitioned from a stationary structure to a speeding, ascending rocket, leaving a trail of bright flame against the backdrop of the launch site. The controlled explosion of fuel within each Raptor engine provided the thrust necessary to lift the multi-ton craft upward, breaking the bonds of gravity in a precisely orchestrated motion. The ground trembled as the engines continued to fire in perfect synchronization, lifting Starship with a force that had been finely calculated for this very moment.

As Starship ascended, the vehicle's control systems maintained stability and direction, correcting for even the slightest variations in wind and atmospheric resistance. This was no simple task—ascending through Earth's atmosphere introduces an array of forces that could disrupt the vehicle's trajectory if not accounted for. SpaceX's engineers had equipped Starship with advanced guidance and control systems, allowing it to adjust itself hundreds of times per second. The vehicle's stability during ascent relied on a delicate balance, as its internal systems constantly adapted to the shifting aerodynamic pressures and gravitational pulls that worked against it.

During this phase, the flight computers worked in real-time to interpret data from multiple sensors distributed throughout the craft. Every millisecond counted; the slightest miscalculation in trajectory or stability could lead to a deviation that would compromise the entire mission. Starship's control systems communicated with the engines, making minor but critical adjustments to maintain a smooth and direct ascent. This stability was achieved through a combination of gyroscopic stabilizers, thrusters, and feedback from external data sources, ensuring that the vehicle remained on its designated flight path.

As the Starship continued its climb, it entered a phase known as Max-Q, the point at which the aerodynamic stress on the vehicle reached its peak. Here, the strength of the rocket's structural components and the precision of its control systems were tested to their limits. The vehicle's internal systems were programmed to throttle back engine power just enough to reduce stress, carefully balancing the need for continued ascent against the pressure exerted by the dense lower atmosphere. As Starship passed through Max-Q, the engines powered back up, and the vehicle resumed its accelerated climb with renewed force.

The initial flight path was designed with meticulous precision, guiding Starship through a trajectory that would conserve fuel while maintaining the necessary altitude and speed for the upcoming stages of the test. Every adjustment, every burst of energy, and every slight change in direction was pre-programmed, calculated to optimize both stability and efficiency. As Starship continued its upward journey, the vehicle's systems worked seamlessly, each component playing its role to ensure the mission stayed on course, inching closer to the next critical milestone in its journey through the sky.

As Starship continued its climb, it encountered a critical phase known as Max-Q—the point where the aerodynamic forces acting upon the vehicle reach their highest intensity. This phase is one of the most challenging moments in any rocket's ascent, as it places extreme stress on the structure and requires an intricate balance between power and control. For Starship, overcoming Max-Q was both a testament to its robust design and a demonstration of the precision that SpaceX engineers have built into the vehicle's systems.

During Max-Q, the vehicle is moving at high speed through the dense lower layers of Earth's atmosphere. The combination of velocity and air density results in maximum pressure on the rocket's exterior. If these forces aren't managed carefully, they could potentially cause structural damage, putting the entire mission at risk.

SpaceX addressed this challenge by programming the Raptor engines to temporarily throttle down during Max-Q. This reduction in thrust alleviates some of the pressure on the vehicle, allowing it to pass through this stressful phase with minimized risk to its structure.

Once Starship cleared Max-Q, the Raptor engines gradually throttled back up, resuming full power and pushing the vehicle further along its planned trajectory. The careful orchestration of these engines is an engineering marvel in itself. Positioned at the base of the Starship, each Raptor engine is meticulously timed to produce the necessary thrust while maintaining overall balance and control. Unlike traditional engines, which burn only liquid fuel, the Raptor engines are powered by a mix of liquid methane and liquid oxygen, a choice that enhances efficiency and supports SpaceX's long-term goal of refueling Starship on Mars.

During ascent, the engines work in unison, firing in specific sequences that adjust as the vehicle gains altitude. The flight computer controls each engine individually, monitoring real-time data and making minute adjustments to maintain the correct flight path. The complexity here is immense: the engines must provide enough thrust to keep the vehicle ascending while simultaneously compensating for any external forces that could pull it off course. This requires a careful orchestration of power and direction that only advanced computing and real-time data processing can achieve.

Together, the control of Max-Q stress and the precisely timed operation of the Raptor engines allow Starship to maintain a stable and efficient trajectory through the atmosphere. This finely tuned balance of force and structural resilience brings Starship closer to its ultimate goals—achieving the milestones set out for the flight and demonstrating the readiness of this advanced technology for the rigorous demands of space travel. With each successful phase, SpaceX moves one step closer to redefining what's possible in reusable rocketry, paving the way for future missions that will push beyond Earth's atmosphere and into the broader cosmos.

Chapter 3: Booster Separation and Boost-Back Burn

As Starship reached a specific altitude and speed, the next critical milestone approached: booster separation. This stage is essential for the reusable design of SpaceX's Starship system, as it marks the moment when the Super Heavy booster, having provided the necessary thrust to lift Starship into the upper atmosphere, detaches from the spacecraft and begins its journey back to Earth. This process is more than just a technical maneuver; it's central to SpaceX's mission of creating a fully reusable launch system that can dramatically reduce the cost of space travel. By reusing the booster, SpaceX can save millions on each launch, making regular space missions economically viable and sustainable.

Booster separation itself is a finely orchestrated process. At the predetermined altitude, when the booster has exhausted the fuel allocated for its portion of the ascent, an automated system initiates the separation. Precision is critical here—the booster must disengage smoothly and with enough distance from Starship to ensure that both the booster and the vehicle continue on their designated paths. Any deviation could impact Starship's ascent or complicate the booster's return trajectory. Once detached, Starship continues its climb toward its destination, while the booster transitions into the next phase of its journey: the boost-back burn.

The boost-back burn is an integral step in returning the Super Heavy booster to its launch site for recovery. After separation, the booster has momentum pushing it away from the site, and this burn helps to reverse its trajectory, guiding it back toward its landing target. Essentially, the boost-back burn involves reigniting a portion of the booster's Raptor engines to slow it down and alter its course, pivoting it around for a controlled descent. This maneuver is both complex and energy-intensive, as it requires precise calculations to overcome the booster's initial trajectory, adjust for altitude and speed, and position it on a path back to Earth.

The timing and accuracy of the boost-back burn are critical for the success of the recovery operation. If the burn is initiated too early or too late, or if the thrust applied isn't accurately controlled, the booster could miss its landing zone. To manage this, SpaceX relies on real-time data from the booster's navigation systems, which constantly monitor its position, speed, and orientation. Adjustments are made on the fly, allowing the booster to follow a precise descent path.

This boost-back burn isn't just about returning the booster; it's also about preparing it for a safe, controlled landing. By aligning the booster with its landing trajectory, SpaceX can guide it through the atmosphere in a way that minimizes wear and tear, preserving it for multiple future launches. This step is essential for the reusability of the booster, as each successful recovery enables SpaceX to inspect, refuel, and prepare the booster for subsequent missions with minimal refurbishment.

In achieving booster separation and executing the boost-back burn, SpaceX demonstrates the sophistication of its reusable rocket technology, setting a new standard in spaceflight. Each element, from precise detachment to trajectory correction, reflects SpaceX's dedication to revolutionizing space travel by creating systems that are not only advanced but sustainable. With every successful recovery, the dream of affordable, routine space missions comes closer to reality, showcasing the power of reusability in reshaping humanity's reach into space.

The success of Starship's fifth flight test relied heavily on the flawless timing and execution of both booster separation and the subsequent boost-back burn. These maneuvers are pivotal in any launch sequence, but their precision becomes even more crucial when reusability is the goal. SpaceX has meticulously designed each phase to unfold in perfect sequence, with split-second timing ensuring that every piece of the puzzle fits together to achieve the broader mission objectives.

Booster separation occurs when the booster has completed its ascent role, providing enough thrust to propel Starship to the required altitude. The timing here is essential; separation has to happen at the precise moment when the booster's fuel allotment has been optimally utilized, maximizing efficiency while avoiding any potential drag on Starship as it continues its ascent. This careful orchestration prevents excess fuel waste and ensures that Starship can continue climbing at the intended velocity and angle. Automated systems trigger the separation sequence, releasing the booster with a precise push that allows it to cleanly and smoothly disengage from Starship. This smooth disengagement is key, as any error in this phase could impact the trajectories of both vehicles, potentially compromising the mission.

Immediately after separation, the boost-back burn becomes the next crucial step. For this maneuver, SpaceX's engineering team has developed advanced technology that enables the booster to perform an almost immediate turnaround, redirecting itself back toward the launch site for a controlled landing. Timing is essential in the boost-back burn as well; any delay or miscalculation could disrupt the booster's descent, potentially missing the targeted landing area or requiring it to use additional fuel, impacting efficiency and overall mission costs. SpaceX's Raptor engines, designed with high efficiency and precision, play a central role here, reigniting on command to provide the necessary reverse thrust that alters the booster's path and slows its initial trajectory.

The technology enabling these precise movements is a testament to SpaceX's engineering prowess. Advanced guidance systems embedded within the booster and Starship continuously monitor variables like altitude, speed, and angle, feeding real-time data to onboard computers that control each engine's thrust and direction. The engines respond dynamically, fine-tuning their output to match the ideal burn rate for both ascent and descent phases. This highly adaptive system allows for on-the-fly adjustments, ensuring that the booster stays within a pre-determined flight corridor that will bring it safely back to the landing site.

One of SpaceX's major innovations is its proprietary software that processes the vast amounts of real-time data coming from sensors embedded across the vehicle. These sensors relay information about temperature, pressure, speed, and positioning, allowing the flight computer to instantly adapt the booster's trajectory and engine output as needed. By merging software with powerful hardware, SpaceX has created a system that performs at a level of precision unheard of in previous reusable rockets.

Beyond the technical precision, the sheer engineering challenge of these maneuvers highlights the innovative spirit driving SpaceX. The successful synchronization of separation and boost-back burn requires an integrated system of sensors, controls, and engines that can handle complex, split-second decision-making without human intervention. Each component, from the Raptor engines to the navigation algorithms, is a marvel of engineering, carefully crafted to handle the immense pressures and temperatures of space travel while maintaining the booster's structural integrity for reuse.

Together, the precision timing, advanced control systems, and cutting-edge technology that SpaceX employs in these operations mark a new era in space exploration. By achieving such accuracy, SpaceX not only completes successful missions but also strengthens the foundation for regular, reliable, and affordable space travel, bringing the vision of a multi-planetary future closer to reality.

Chapter 4: The High-Stakes Catch Decision

In the intense and fast-paced environment of a SpaceX launch, the role of the flight director and their team becomes essential, especially during critical moments like booster catch decisions. The flight director, serving as the nerve center of mission control, is responsible for coordinating the vast network of specialists and systems, each of whom provides real-time data on different aspects of the mission. The flight director must make quick, decisive calls, relying on their team's expertise and the accuracy of incoming data to assess whether the conditions are right for complex maneuvers, such as the booster catch.

In the seconds leading up to the booster's return, the flight director must evaluate a series of critical factors. Wind speed, descent angle, engine thrust, and alignment with the landing zone all play a part in determining whether the booster can attempt a safe catch or if an alternative approach is needed. These decisions happen in real time, with the team responding dynamically to conditions that may shift within moments. A successful catch doesn't just depend on the engineering of the vehicle but on the calculated decisions made by the flight director and their team in mission control, who orchestrate each phase of the return with precision.

A key element that adds complexity to these decisions is the hot-stage separation, a maneuver that enables the booster to initiate its return while Starship continues its ascent. Hot-stage separation, unlike traditional separation where all engines are shut off, involves briefly igniting the next stage's engines while the first stage is still active, allowing for a smooth, energy-efficient transfer of propulsion from the booster to Starship. This technique is challenging to execute, but it minimizes time lost during separation and preserves momentum for both stages, improving the overall efficiency of the launch.

For reusability, the hot-stage separation is a game-changer. By allowing the booster to separate and begin its return journey sooner, it conserves fuel that would otherwise be used to counteract additional descent speed or to adjust its trajectory later. This early maneuver enables the booster to align more accurately with its landing path, reducing the complexity of the boost-back burn and saving energy for the final descent. With this level of fuel conservation and efficiency, SpaceX can achieve more sustainable reusability, extending the potential lifespan of each booster and allowing for more frequent launches without significant refurbishments.

The hot-stage separation places high demands on both the vehicle's structure and its control systems. As the booster and Starship separate, they are still exposed to intense forces, and any instability could affect the trajectory of both stages. SpaceX's engineering innovations in guidance and stability systems, combined with the expertise of the flight director and the mission control team, allow them to manage these risks effectively, preserving the booster for a controlled descent and planned catch.

By combining real-time decision-making with innovative flight maneuvers like hot-stage separation, SpaceX continues to redefine the technical limits of reusable rocketry. Each successful catch and return isn't just a technical feat; it's the result of a well-coordinated team effort, guided by the knowledge and expertise of those behind the scenes, who work together to ensure that each mission milestone brings the vision of sustainable, repeatable space travel closer to reality.

In the moments leading up to the final catch attempt, the atmosphere in mission control becomes charged with a heightened sense of focus and intensity. The countdown to the catch begins, and the clock ticks down the last 30 seconds—a crucial window during which the flight director and their team assess every aspect of the booster's descent, calculating whether conditions are optimal for the catch attempt. It's a period when the entire mission hangs in the balance, with each second bringing them closer to a decision that could mark either a successful recovery or a missed opportunity.

The flight director, fully aware of the stakes, has their team closely monitoring a variety of critical metrics in real time. They analyze descent speed, wind conditions, fuel levels, and the precise positioning of the booster relative to the landing zone. The data floods in from a network of sensors, all feeding into mission control's systems, where every variable is tracked down to the smallest detail. The booster's Raptor engines, vital for making last-minute adjustments, must be ready to respond instantaneously, and the "Chopsticks" catching mechanism must be perfectly aligned to intercept the booster at exactly the right moment.

This narrow window presents immense challenges. Environmental factors like wind gusts or atmospheric turbulence can shift unexpectedly, even within the final few seconds, affecting the booster's descent path. Any misalignment between the booster and the landing structure would force the flight director to make the difficult decision to abort the catch and allow the booster to land elsewhere, prioritizing the safety and integrity of both the vehicle and ground systems. This split-second decision-making requires absolute concentration, with the flight director needing to balance the booster's trajectory against real-time conditions to ensure the success and safety of the operation.

The risks are high. If the booster's descent speed is too rapid, or if a sudden change in trajectory places it outside the optimal catch range, attempting to intercept it could lead to structural stress on the catching mechanism or potential damage to the booster itself. The team's calculations and decisions must be flawless, with no margin for error. The final seconds demand not only technical precision but also judgment and experience, qualities that the team and flight director bring to the mission from countless hours of preparation and previous launches.

As the countdown reaches its final seconds, all systems are "go." The booster lines up with the "Chopsticks," descending toward its designated landing point. The tension peaks as the flight director gives the final confirmation, clearing the attempt. With pinpoint timing, the booster is caught and secured—a testament to the seamless orchestration of technology and human skill under pressure.

In the live mission environment, every decision carries weight. Each team member, from the flight director to the data analysts, knows that their actions contribute to the success or failure of this mission-critical phase. It's a scenario where technology and human expertise converge, with both playing an indispensable role in making reusability a reality. This ability to execute complex maneuvers in real time sets SpaceX apart, demonstrating the resilience and innovation that are redefining the boundaries of what's possible in space exploration. The final catch, a triumph of calculated precision, underscores the level of mastery required to achieve reliable and repeatable spaceflight.

Chapter 5: The Booster Return and Catch Attempt

As the Super Heavy booster began its return, its trajectory and descent were carefully controlled, guided by precise calculations and a series of critical maneuvers. From its peak altitude, the booster arced gracefully, rotating to face the direction of its landing zone. This turn marks the beginning of its descent, a controlled fall back to Earth. Every adjustment to its path was calculated to bring it into alignment with the landing target, maximizing efficiency while minimizing the distance it needed to cover on its way back.

The view from mission control was a mix of data streams and live feeds, each offering a unique perspective on the booster's journey home.

Monitors displayed graphics tracking the booster's path, overlaying its trajectory on real-time data to visualize its distance from the landing site, speed, angle, and expected time of arrival. Engineers and specialists followed each reading intently, noting any deviations and assessing how changes in speed or position might affect the descent. On some screens, a live video feed captured the booster as it began its re-entry into the atmosphere, showing the gleaming metallic body surrounded by the heat and turbulence of re-entry.

As the booster dropped toward its targeted landing area, the Raptor engines ignited to initiate a controlled deceleration. This ignition marked a critical stage in the descent, as these engines provided the power needed to slow the vehicle to a safe speed and stabilize its orientation. The view from mission control shifted as telemetry data reflected the sudden deceleration, showing the effects of the Raptor engines as they kicked in. With each adjustment, the booster became more aligned with the pre-planned descent path, the engines working to fine-tune its angle and speed, allowing it to approach the landing zone with precision.

The booster's descent and the re-ignition of the Raptor engines were more than just technical maneuvers; they were a testament to SpaceX's engineering and coordination in action. The live video feed showed flames and exhaust streaming from the engines as they fired in unison, slowing the booster in the final moments of descent. From mission control, the team watched in anticipation as the booster continued to descend, each second bringing it closer to the dramatic conclusion of its journey. Each display and monitor showed a different aspect of the descent, from altitude and velocity to fuel reserves, providing the team with all the data they needed to monitor the descent with absolute precision.

As the booster approached its target, the Raptor engines adjusted yet again, ensuring that the vehicle's speed and position were within the range required for a successful catch. The carefully controlled burn allowed it to settle directly over the landing site, where it hovered momentarily before the final touchdown. In the mission control room, the tension gave way to a quiet satisfaction as the screens confirmed the booster had returned successfully, marking another triumph for SpaceX in its mission to make reusable rockets a reality. This return, captured and analyzed in real-time, highlighted the incredible orchestration of technology, expertise, and timing that made the catch a success, bringing SpaceX one step closer to sustainable and frequent space travel.

As the Super Heavy booster neared its landing site, it entered the most critical stage of its descent: the landing burn phase. This final burn is vital for slowing the booster to a safe landing speed, reducing its rapid descent to a controlled approach. Precision is paramount here, as any miscalculation in speed, angle, or timing could jeopardize the entire recovery. The booster's Raptor engines, which had been carefully monitored throughout the return, ignited once again, providing just enough thrust to decelerate and stabilize the vehicle as it moved into position over the designated landing zone.

In a groundbreaking maneuver, SpaceX employed its innovative "Chopsticks" mechanism—a set of massive, robotic arms engineered to catch the returning booster. This mechanism represented an evolution in booster recovery, designed to "catch" rather than land the booster, minimizing wear on the structure and allowing for faster turnaround and reuse. Positioned on the landing tower, the Chopsticks were programmed to open and align themselves with the descending booster, ready to intercept it with precision timing.

From mission control, the view of the landing was a breathtaking display of engineering in action. The team watched the booster's slow, steady descent as it aligned with the Chopsticks, guided by onboard sensors and control algorithms. Each thrust adjustment from the Raptor engines was carefully executed to keep the booster centered and stable, easing it toward the arms that awaited it. Data monitors in mission control displayed real-time calculations on speed, position, and alignment, each figure inching closer to the exact parameters required for a successful catch.

In those final moments, the tension in mission control was tangible, as all eyes were on the screens displaying the booster's approach. As the booster descended into the arms of the Chopsticks, there was a collective anticipation—years of innovation and planning, countless hours of testing, and the expertise of hundreds of engineers had all led to this singular moment. The Chopsticks moved with precise synchronization, closing around the booster with a controlled grip. When the booster came to a secure halt, there was a shared recognition that history had just been made.

The successful catch of the Super Heavy booster was more than an engineering feat; it was a testament to teamwork and the ambitious vision that defines SpaceX. Each member of the team, from those who designed the mechanisms to those who monitored each data point in mission control, had contributed to this achievement. It was a moment that demonstrated the power of collaboration, where every individual effort came together to push the boundaries of space exploration.

This historic catch not only showcased SpaceX's commitment to reusability but also set a new standard in rocket recovery. By catching the booster instead of landing it conventionally, SpaceX reduced the impact and potential wear, preserving the integrity of the booster for future flights. The innovation behind the Chopsticks mechanism and the successful execution of the landing burn phase reflect a level of mastery in reusable rocketry that brings the vision of affordable, routine space travel closer to reality. As the booster hung securely in the arms of the Chopsticks, it became clear that this was more than just a successful test—it was a milestone in humanity's journey toward exploring the cosmos.

Chapter 6: Starship's Journey Beyond Booster Separation

With the Super Heavy booster successfully separated and on its way back to Earth, Starship continued its ascent, driven by its own Raptor engines toward the edge of space. This phase marked the vehicle's transition from the Earth's atmosphere into the vacuum of space, a journey where precise control and optimized fuel usage were crucial. Starship's flight path was carefully planned to maximize efficiency, allowing it to conserve as much energy as possible for its main objective and eventual return. As it soared higher, the Earth below began to recede, and the vehicle entered the weightlessness of space—an environment free from the atmospheric resistance that had once worked against it.

This transition was significant, as it demonstrated Starship's ability to operate autonomously in space, navigating without the support of atmospheric lift. The vehicle's guidance systems continued to monitor its trajectory, while advanced propulsion mechanisms kept it on course, reaching the target altitude necessary for its mission. SpaceX had designed Starship with a streamlined exterior and heat-resistant materials to prepare it for the unique challenges of space travel and, most critically, for its eventual re-entry into Earth's atmosphere. The ascent was smooth and controlled, highlighting SpaceX's expertise in long-duration engine performance and the craft's resilience against the intense forces encountered at high altitudes.

As Starship completed its mission objective and began its descent, it prepared to face one of the most daunting phases of its journey: the peak heating stage of re-entry. Returning from space at tremendous speed, Starship encountered Earth's thickening atmosphere, causing rapid compression of air around the vehicle and generating extreme temperatures on its surface. This phase is one of the greatest tests of any spacecraft, as the thermal energy generated by re-entry can reach thousands of degrees, capable of compromising the integrity of the vehicle if not properly managed.

To withstand this intense heat, SpaceX equipped Starship with advanced thermal protection systems. The vehicle's exterior was lined with heat-resistant tiles, strategically designed to absorb and deflect the extreme thermal energy created during re-entry. This protection system had been rigorously tested and optimized, as even a single flaw could expose the spacecraft to dangerous temperatures, risking damage to its structure. The heat shield tiles worked in tandem with the vehicle's aerodynamic design, which minimized friction while guiding the vehicle through its controlled descent angle.

During the peak heating phase, Starship's sensors continuously relayed data back to mission control, monitoring temperatures, structural integrity, and the vehicle's alignment. The on-board systems adjusted its descent path and attitude to optimize thermal protection, ensuring that the most heat-resistant sections of the craft bore the brunt of re-entry stress. These real-time adjustments were critical in preventing overheating and stabilizing the vehicle's descent.

Successfully navigating the peak heating phase required more than just advanced materials—it demanded a precise balance between engineering and real-time adaptability. SpaceX's design allowed Starship to maintain a stable trajectory while dispersing heat effectively, preserving its structural integrity for the final stages of descent. This capability, showcased in Starship's seamless transition through re-entry, underscored the sophistication of SpaceX's engineering, reflecting the company's commitment to developing reusable spacecraft that can endure the intense demands of repeated space travel.

With the worst of re-entry behind it, Starship's journey was nearly complete. The vehicle had proven itself against one of the most extreme environments it would ever face, emerging from the searing heat intact and ready for the final landing sequence. Each phase of its descent, from controlled entry to peak heating management, demonstrated a breakthrough in reusable rocketry, paving the way for future missions that could push the boundaries of human exploration even further.

The significance of Starship's reusability extends far beyond just recovering its booster; it represents a fundamental shift in how humanity approaches space exploration. Traditional rockets, used only once before being discarded, impose high costs and logistical limits on each mission. With Starship, however, SpaceX has designed a system where both the booster and the spacecraft itself can be recovered, refurbished, and flown again—bringing down the cost per launch dramatically. This capability not only changes the economics of spaceflight but also unlocks a range of possibilities that previously seemed like distant dreams.

Reusability enables Starship to support high-frequency launches, which is essential for ambitious projects such as building permanent human settlements on Mars or establishing lunar bases. By reusing the same vehicle multiple times, SpaceX can conduct missions that would otherwise be prohibitively expensive or logistically overwhelming. Starship could act as a supply vessel, ferrying large quantities of cargo and equipment to support infrastructure in space or on other planetary bodies. The ability to carry heavy payloads repeatedly without the cost of building a new rocket for each mission means that Starship could one day establish a consistent "bridge" between Earth and distant destinations, making continuous support of interplanetary colonies a reality.

Beyond the Moon and Mars, Starship's capabilities open doors for deep-space exploration. With its substantial payload capacity and versatility, it could carry large scientific instruments or even entire research modules to explore outer planets, asteroid belts, or far-off destinations like Jupiter's moons. The fact that Starship is designed to refuel in space—another aspect of its reusability—adds to its potential. By docking with an orbiting fuel depot, Starship could extend its range significantly, reaching locations farther than any human has traveled before. This refueling capability is essential for creating a sustainable framework for space exploration, where vehicles can operate far beyond Earth's orbit without constantly returning for resources.

Closer to home, reusability transforms space into an accessible frontier for industries that previously lacked feasible access to orbit. Starship's capacity to fly frequently and at reduced cost has enormous implications for commercial satellite deployment, space tourism, and scientific research. Satellite companies could deploy large constellations to improve global communications, researchers could conduct experiments in orbit more affordably, and space tourism companies could bring the dream of experiencing weightlessness or viewing Earth from above to more people. By drastically reducing the price barrier, Starship allows a greater diversity of participants in space activities, fostering an ecosystem where new industries, innovations, and partnerships could thrive.

On Earth, the impact of Starship's reusability resonates through its influence on environmental sustainability and technological advancements. By using a fully reusable system, SpaceX minimizes waste and conserves resources, setting a new standard in aerospace for eco-friendly operations. The drive to perfect reusable rockets has also led to technological breakthroughs that could benefit other fields, from advanced materials science to artificial intelligence and machine learning applications in real-time data processing and autonomous control.

Starship's vision of reusability symbolizes a new era in human spaceflight. Each successful launch, recovery, and relaunch reinforces the possibility of a future where access to space is as regular as air travel, and where humanity can genuinely contemplate life beyond Earth. By achieving these capabilities, SpaceX is not merely building a spacecraft; it is laying the groundwork for humanity's transition into a multi-planetary species, bringing the extraordinary within reach and transforming the unimaginable into the attainable. The potential of Starship is not just about reaching the stars; it's about reshaping the very nature of human existence, opening pathways that lead us further into the universe and closer to understanding our place within it.

Chapter 7: Atmospheric Re-Entry and Heat Management

As Starship re-enters Earth's atmosphere, it faces one of the most extreme challenges of space travel: the intense heat and friction generated by its high-speed descent. Re-entry occurs at speeds approaching several kilometers per second, and as the vehicle pushes against the dense atmosphere, the air in front of it compresses rapidly, creating an immense buildup of thermal energy. This compression can cause temperatures on the vehicle's surface to soar to thousands of degrees—enough to melt metal and weaken structural components if left unchecked. The heat during re-entry is a formidable obstacle, one that has historically made space vehicles difficult to recover intact, let alone reuse.

SpaceX has approached this challenge with a combination of advanced materials, meticulous design, and innovative engineering. To protect Starship during re-entry, SpaceX has outfitted it with a specialized thermal protection system consisting of heat-resistant tiles. These tiles, made from materials designed to withstand extreme temperatures, cover the spacecraft's surface, particularly on the side that faces forward during re-entry. This protective "shell" serves as a shield, absorbing and deflecting the intense thermal energy generated by the friction with the atmosphere.

Each heat shield tile is meticulously crafted and installed to ensure complete coverage and resilience. The materials used are lightweight yet robust, designed to handle the rapid heating and cooling cycles Starship encounters. The tiles are arranged to create an insulating layer that can absorb and radiate the heat away from the vehicle's main structure, preventing it from sustaining damage that could compromise its integrity. SpaceX has also rigorously tested these tiles, refining them through numerous test flights to identify and address any vulnerabilities, ensuring they can perform consistently under the intense conditions of re-entry.

The shape of Starship itself contributes to its ability to endure peak re-entry temperatures. With its aerodynamic design, SpaceX has configured Starship to orient itself at a specific angle during descent, allowing it to "belly-flop" through the atmosphere. This unique entry method maximizes the surface area exposed to atmospheric friction, which distributes the heat more evenly across the heat shield tiles and reduces the stress on any single point. The belly-flop maneuver also helps to slow Starship down naturally, allowing for a more controlled descent without over-reliance on its engines to decelerate.

Throughout re-entry, sensors embedded across Starship's heat shield monitor temperature fluctuations in real-time, providing data that the onboard systems use to make adjustments to the vehicle's orientation and descent angle. This real-time feedback loop helps SpaceX engineers assess how the heat shield performs under pressure, allowing the system to adapt to changing conditions. The data collected from each re-entry is invaluable, as it helps SpaceX refine Starship's design to further enhance its resilience against the severe conditions of re-entry.

The combination of heat-resistant materials, aerodynamic design, and real-time adaptability makes Starship a groundbreaking vessel capable of withstanding the punishing demands of re-entry, setting a new standard in reusable spacecraft technology. SpaceX's design not only ensures the vehicle's safety but also preserves its structural integrity for multiple flights, reinforcing its role as a truly reusable spacecraft. Each successful re-entry confirms Starship's ability to endure and recover, pushing forward the boundaries of what's possible in space exploration and bringing humanity closer to routine, sustainable access to space.

As Starship begins its descent back into Earth's atmosphere, it transitions from the weightlessness of space into the dense, friction-laden layers of air surrounding the planet. This atmospheric entry phase is carefully orchestrated to manage the intense heat and forces that Starship encounters during its return. Each step in this sequence is critical, balancing precision with the robustness required to withstand some of the most extreme conditions a spacecraft can face.

The entry process starts with Starship orienting itself for a controlled descent. Rather than nose-diving straight down, Starship performs a "belly-flop" maneuver, positioning its broad, flat side toward the atmosphere. This orientation maximizes the surface area exposed to atmospheric friction, which helps distribute the intense heat generated during re-entry more evenly across its heat shield. The maneuver also serves to slow Starship naturally, allowing gravity and atmospheric drag to work in tandem to reduce its speed before it reaches lower altitudes where controlled burns will take over.

As the vehicle accelerates through the upper atmosphere, the heat-resistant tiles on Starship's belly come into play. These tiles, crafted from advanced materials specifically chosen for their ability to withstand the high temperatures of re-entry, act as a protective barrier against the searing heat caused by rapid air compression. Each tile is designed to absorb and deflect thermal energy away from Starship's main structure, maintaining its integrity and protecting its components from overheating. Arranged in a seamless, overlapping pattern, these tiles ensure that no part of Starship's skin is left vulnerable to the atmospheric friction encountered during descent.

In addition to the tiles, Starship's design incorporates innovations in heat management that help it handle the fluctuating temperatures experienced during this stage. Sensors distributed across the heat shield measure real-time temperature changes and structural integrity, relaying this data back to Starship's onboard systems. These sensors not only monitor the heat levels but also track any minor shifts or pressures exerted on the shield, enabling the system to make micro-adjustments to Starship's angle of descent as needed. This adaptability is vital; if certain areas of the shield show signs of heightened stress, the onboard systems can shift Starship's position slightly to redirect some of that pressure, balancing the heat load to prevent potential damage.

As Starship descends, it continues to encounter increasing atmospheric density, further escalating the thermal load on its exterior. The advanced design of the heat shield tiles allows them to endure repeated heating cycles, so even as the temperature spikes, the tiles do not warp or degrade. The materials used have undergone rigorous testing to withstand these conditions, ensuring that they can handle multiple re-entries without compromising Starship's overall reusability. SpaceX has designed this system for endurance, knowing that each mission demands that the shield perform flawlessly not only to protect the vehicle but also to maintain the vision of a truly reusable spacecraft.

The belly-flop descent continues as Starship reaches lower altitudes. By maintaining a horizontal orientation, it maximizes drag and keeps its descent controlled, relying on natural atmospheric resistance to slow it down. The vehicle's body, built to withstand aerodynamic stress, continues to absorb and distribute the intense forces that push against it. Finally, as Starship approaches the designated altitude for controlled landing, it reorients itself, moving from the broadside belly-flop position to a vertical stance for the landing burn sequence.

At this stage, the heat shield tiles have completed their work, successfully guiding Starship through the fiery re-entry phase without compromising structural integrity. The vehicle's resilient design, coupled with real-time adaptability and a robust heat management system, enables it to endure and recover intact. This approach to atmospheric entry marks a significant innovation, as it not only prioritizes Starship's safety but also supports its reusability, allowing it to be quickly refurbished and ready for another mission.

SpaceX's advancements in heat management and structural durability during descent reflect a new chapter in spacecraft design. Each re-entry cycle reinforces Starship's capability to withstand intense environmental challenges, a critical factor in SpaceX's mission to establish regular, cost-effective access to space. With each successful descent, Starship proves its worth as a reliable, reusable vehicle that can weather the demands of space travel and return ready for future journeys. This carefully engineered balance of materials, design, and real-time control is what makes Starship not just a spacecraft, but a sustainable bridge to our exploration of the universe.

Chapter 8: Final Descent and Targeted Landing

As Starship approaches the final phase of its descent, the transition from freefall to controlled landing demands impeccable precision. After withstanding the intense heating and stress of re-entry, the vehicle must now switch from its belly-flop orientation to an upright position for a safe touchdown. This critical maneuver sets the stage for the landing burn, a controlled ignition of its Raptor engines, designed to decelerate the vehicle and guide it smoothly to the landing site.

The landing burn sequence begins with Starship rotating to a vertical position, aligning its base toward the ground. At this stage, every aspect of the descent, from speed to trajectory, must be meticulously controlled. The landing burn ignites the Raptor engines just as the vehicle reaches its designated altitude for final approach. These engines, built for both power and precision, throttle up to provide the necessary thrust to counteract Starship's momentum, slowing its descent to a manageable speed. The goal is to bring the vehicle's downward velocity down to nearly zero as it nears the surface, allowing it to settle safely onto the landing pad.

This delicate operation requires the coordination of complex landing software, which plays a central role in managing every variable in real time. The landing software monitors inputs from an array of sensors across the vehicle, which measure altitude, speed, orientation, and proximity to the landing zone. These sensors relay data to the flight computer, which processes it almost instantaneously to adjust the engine's thrust and angle as needed. The software's real-time adjustments are crucial for achieving the precise control necessary for a safe landing, particularly when dealing with potential variations in atmospheric pressure, wind conditions, or minor shifts in descent angle that could otherwise disrupt a perfect landing.

The landing software's real-time adaptability is one of the most impressive features of Starship's design. The system continuously recalculates Starship's descent trajectory, adjusting engine power to keep it within the precise speed and orientation required for a stable approach. This responsiveness ensures that the vehicle can adapt to subtle changes in its environment, compensating for any deviations to maintain alignment with the landing pad. As the Raptor engines adjust their thrust to fine-tune the descent, Starship maintains an upright position, minimizing lateral drift and aligning perfectly with its touchdown target.

These final seconds of descent are a display of calculated precision and split-second control. The Raptor engines continue to throttle down as Starship nears the surface, reducing its speed to just the right level to ensure a soft landing. The landing software's ability to make real-time adjustments, even in these final moments, exemplifies the advanced engineering behind SpaceX's reusable rocket technology.

The success of Starship's landing burn is a result of this synergy between powerful engines, adaptive software, and precise control mechanisms, each element working in harmony to bring Starship home safely. As the engines shut down and the vehicle settles onto the landing pad, it marks not only the conclusion of a single journey but also another step forward in the evolution of reusable rocketry. Each successful landing reinforces the practicality and reliability of SpaceX's approach, proving that a spacecraft can not only reach the stars but return to Earth, ready to launch again. This cycle of controlled launches, re-entries, and landings showcases SpaceX's commitment to sustainable space travel, opening the door to a future where the journey to and from space is as routine as any other mode of transportation.

In the final seconds of Starship's descent, as it hovered above the landing pad, every adjustment, calculation, and maneuver led to this climactic moment. The Raptor engines, which had been expertly throttled to control speed and direction, now reduced their thrust to just enough for a gentle descent, stabilizing the massive spacecraft for a flawless touchdown.

Starship, having endured the trials of launch, re-entry, and precise landing maneuvers, settled smoothly onto the pad, marking the end of its mission with the precision of a well-executed ballet.

Inside SpaceX's mission control, the atmosphere shifted from intense focus to palpable relief and exhilaration. Engineers, operators, and technicians who had been monitoring every stage of the mission in real time were now witnessing the culmination of their collective efforts. The successful touchdown wasn't just the completion of one flight; it was a testament to the countless hours of planning, simulation, and testing, all aimed at achieving the reusable spaceflight vision that SpaceX had championed from the start.

Almost immediately after landing, the post-mission assessments began. Engineers monitored data from the vehicle's sensors, checking for any anomalies that may have arisen during descent and landing. Heat shield integrity, fuel levels, engine performance, and structural condition were all meticulously examined to ensure that the spacecraft had completed its journey without sustaining damage. Each of these assessments provided invaluable feedback, helping the team refine and enhance future missions. The information gathered in these moments of post-landing analysis contributes to SpaceX's ongoing commitment to improving the safety, reliability, and reusability of Starship.

As the data confirmed the vehicle's successful landing and overall health, the mission control room erupted into celebration. Cheers, applause, and smiles filled the room as the team recognized their achievement. This wasn't just a victory for SpaceX; it was a milestone in space exploration, a testament to human ingenuity and the power of innovation. Each member of the team, from the engineers and software developers to the ground crew and flight director, shared in the triumph of knowing they had taken one more step toward redefining what was possible in space travel.

The successful landing of Starship wasn't merely a technical accomplishment; it was a moment of inspiration, symbolizing the potential for humanity to reach beyond Earth's boundaries with the confidence that we can return safely. With each mission that ends in a controlled landing, SpaceX inches closer to making space exploration routine, sustainable, and accessible, turning the dream of interplanetary travel into an achievable reality.

As the post-mission celebrations wound down, the SpaceX team already had their eyes on the future. Each successful flight and safe return brings them closer to the vision of regular, affordable journeys into space, and this landing was one more milestone on that path. For now, though, they could savor the success of this mission, a moment where everything—from complex engineering to real-time decision-making—came together to bring Starship safely back to Earth, ready for its next adventure among the stars.

Chapter 9: Engineering Triumphs and Lessons Learned

The fifth flight test of Starship showcased a series of groundbreaking engineering innovations, pushing the boundaries of reusable rocketry and bringing SpaceX closer to realizing its vision of affordable, sustainable space travel. At every stage, from launch to landing, the mission highlighted advancements that underscored the robustness and adaptability of the Starship system, reflecting years of focused research, development, and testing.

One of the standout innovations was SpaceX's successful use of the "Chopsticks" mechanism to catch the Super Heavy booster on its descent. Rather than relying on conventional landing legs, SpaceX designed massive robotic arms capable of intercepting the booster midair, reducing the impact force on the structure and paving the way for a quicker turnaround. This approach not only preserved the booster's integrity but also represented a critical step toward rapid reusability, as a booster caught in this manner requires less refurbishment between flights. The "Chopsticks" system is a clear example of SpaceX's inventive approach to solving complex engineering problems, using unconventional designs to achieve efficiency and longevity.

Additionally, the precision of the landing burn sequence demonstrated remarkable improvements in real-time guidance and control systems. The ability to throttle Raptor engines precisely, allowing Starship to maintain a stable descent speed and align with the landing pad, was no small feat. This level of control was made possible by advanced landing software, which leveraged data from a network of sensors to make split-second adjustments in real time. These improvements in software and engine control have broad implications for SpaceX's goal of creating a fully reusable space vehicle, providing a level of precision that minimizes risk during landing and enables safe, consistent returns to Earth.

The innovations in thermal protection were equally significant. Starship's heat shield, designed to withstand the extreme temperatures of re-entry, functioned as intended, preserving the vehicle's structure despite the intense frictional heating. The advanced heat-resistant tiles and sensor-embedded systems provided real-time feedback, allowing the vehicle to adjust its orientation slightly to ensure even heat distribution. This capability not only kept Starship safe during this mission but also validated SpaceX's approach to building a heat shield that could endure repeated re-entries with minimal maintenance, a crucial factor in achieving regular, reusable spaceflight.

Each of these engineering feats underscored SpaceX's unique approach to iterative testing and improvement. The mission also provided a wealth of insights, each contributing to future development. First and foremost, the importance of timing and real-time adaptability was reaffirmed. From the intricacies of the go/no-go poll to the timing of the landing burn, every phase highlighted how precise, data-driven decision-making is essential to mission success. Each test provides SpaceX with a clearer understanding of how to refine and optimize these processes, ultimately improving the reliability and efficiency of Starship's systems.

The mission also illuminated areas where enhancements could further support SpaceX's objectives. While the Chopsticks catch mechanism proved effective, its first use offered key insights into fine-tuning the system for smoother, faster catches. Similarly, the landing software's performance will continue to be refined, making adjustments even more responsive to real-time data for optimal safety and stability during landing. With each flight, SpaceX deepens its understanding of these systems and identifies even the smallest areas where incremental improvements can yield significant results.

This fifth flight test marked a defining moment, not only for SpaceX but also for the broader vision of reusable rocketry. Each lesson learned and every successful maneuver contribute to a growing foundation of knowledge and capability that will shape future Starship missions. As SpaceX continues to build on these advancements, the company moves closer to a future where regular, reliable access to space is a practical reality, enabling missions that range from lunar landings to interplanetary exploration. The insights gained from this mission will undoubtedly influence not only the next launch but also SpaceX's long-term roadmap, bringing the dream of sustainable space travel closer with each successful test.

Each stage of Starship's fifth flight test played a critical role in achieving the mission's overarching goals and demonstrated the technical prowess required to bring reusable rocketry closer to reality. The success of the mission was the result of a series of interconnected processes, each dependent on the flawless execution of the one before it. From pre-flight checks to the final catch of the booster, every stage showcased SpaceX's commitment to precision, adaptability, and innovation.

The launch phase set the tone for the mission, with the synchronized firing of Starship's powerful Raptor engines lifting the vehicle from the pad with immense thrust and stability. This initial stage was critical not only for achieving the altitude needed for booster separation but also for testing the engines' reliability and performance under intense conditions. Engineers had to be certain that the engines could deliver consistent thrust and control, allowing the vehicle to reach its precise trajectory without deviation. The success of the launch was a testament to SpaceX's rigorous design and testing protocols, which ensured that every element functioned as intended.

Booster separation and the subsequent boost-back burn were pivotal in demonstrating the viability of SpaceX's reusability goals. Once the booster detached, its boost-back burn required pinpoint accuracy to realign it with the landing site and guide it through a controlled descent. This stage highlighted SpaceX's advanced guidance systems, which made real-time adjustments to compensate for any unexpected variables. Successful booster recovery is foundational to the concept of a fully reusable launch system, and the boost-back burn was a key factor in achieving a safe, controlled return.

The descent phase tested both the booster's structural integrity and SpaceX's innovative Chopsticks mechanism for catching it mid-air. As the booster approached the landing zone, the precisely timed landing burn brought it to a near-halt, allowing the Chopsticks to secure it gently and safely. This step was not only groundbreaking but also a clear validation of SpaceX's ability to handle complex landing procedures that minimize wear on the booster, which is essential for quick turnarounds and multiple reuses. The successful catch underscored the commitment of SpaceX's engineers to innovate beyond conventional recovery methods, showing that rapid reusability is within reach.

For Starship, the re-entry and landing stages demanded equally rigorous testing of heat management and stability controls. The heat shield and thermal protection system were essential to withstand the intense temperatures encountered during re-entry, and their performance demonstrated that SpaceX's design could endure repeated flights. The controlled landing burn, combined with the vehicle's aerodynamic control during descent, underscored the importance of real-time software adjustments and advanced sensor feedback. Together, these technologies ensured that Starship could land safely, marking yet another step toward making regular reusability a reality.

Reflecting on the significance of this test, SpaceX engineers and mission control team members expressed a sense of accomplishment and pride. They recognized that each success brought them closer to SpaceX's larger vision of making space travel affordable and sustainable. For many on the team, this mission was the culmination of years of hard work and incremental improvements, proving that SpaceX's iterative approach could yield results that push the limits of space technology. Engineers in mission control noted that watching the vehicle they had helped design, build, and test succeed in each phase was immensely rewarding, underscoring the importance of collaboration and innovation in achieving such ambitious goals.

The SpaceX team also saw this test as a milestone with profound implications for the future. With each successful recovery, the dream of establishing human settlements beyond Earth feels less like science fiction and more like an achievable reality. Team members expressed excitement for what lies ahead, envisioning missions to the Moon, Mars, and beyond, all made possible by Starship's reusability. Reflecting on the journey from the drawing board to a live test, the team saw each step of the process as part of a larger narrative—one that will shape the future of human exploration and redefine humanity's place in the cosmos.

For SpaceX, the fifth flight test of Starship was not merely a technical achievement; it was a step forward in the collective journey of human space exploration. Each stage brought new insights and reinforced the dedication to building a spacecraft that could change the way we think about reaching and inhabiting space. In the reflections of SpaceX's engineers, the significance of the test became clear: it was a validation of both vision and capability, a moment when the barriers between dreams and reality grew thinner, and the possibility of a multi-planetary future moved closer to realization.

Chapter 10: The Road Ahead for SpaceX and Starship

SpaceX's vision for Starship extends far beyond individual launches, with plans that aim to redefine humanity's relationship with space. In the near term, SpaceX is preparing to refine Starship's capabilities, building on successes from previous flights. A significant focus remains on enhancing the "Chopsticks" mechanism used to catch the returning Super Heavy booster, a technique crucial to achieving rapid reusability. By perfecting this mechanism, SpaceX aims to reduce the wear on the booster, allowing for quicker turnaround times and supporting a sustainable launch cadence.

Looking further ahead, SpaceX has set ambitious goals for Mars exploration, with plans to send uncrewed Starships to the Red Planet within the next few years and potentially launch crewed missions shortly afterward. This vision supports the long-term goal of establishing a self-sustaining city on Mars within the next two decades, a bold step in making life multiplanetary. With regular, reliable access to Mars, SpaceX envisions a future where interplanetary travel and settlement become feasible, shaping the possibility of a human presence on other worlds.

The pursuit of reusability brings evolving challenges that SpaceX is addressing through innovative solutions. One of the most crucial is the development of in-space refueling technologies, which would enable Starship to refuel in orbit, significantly extending its mission capabilities and range. For missions beyond low Earth orbit, such as those to the Moon and Mars, the ability to refuel in space is essential. SpaceX has made strides in propellant transfer technology, allowing Starship to conduct extended missions without needing to return to Earth for refueling.

Economic factors also play a significant role in the reusability equation. While reusability has the potential to lower launch costs drastically, the upfront development expenses are considerable. Additionally, efficient refurbishment processes between flights are essential for realizing the full economic benefit. SpaceX's goal is to reduce the cost per launch to a fraction of traditional expendable rockets, aiming for a price point that could make regular space access affordable and economically viable.

Technical challenges also continue to shape the path forward for reusability. Designing materials and systems that can withstand multiple launches and re-entries without degradation is a complex engineering feat. SpaceX has been investing in advanced materials and structural designs to increase Starship's resilience, ensuring its components can endure the stresses of repeated space travel.

Overall, SpaceX's plans for Starship reflect a transformative approach to space exploration, where reusability and cost-efficiency open new horizons for humanity. While the journey toward full reusability involves tackling technical, economic, and logistical hurdles, each advancement brings us closer to a future where regular space travel, exploration, and settlement become attainable. The efforts underway with Starship are paving the way for a new era of space exploration, where humanity's reach can extend farther and faster than ever before, turning dreams of a multi-planetary existence into reality.

The potential of SpaceX's Starship to transform Mars missions, lunar exploration, and deep-space endeavors marks a profound shift in humanity's approach to space travel. As a fully reusable spacecraft with high payload capacity, Starship is uniquely positioned to support sustained missions to both the Moon and Mars. Unlike previous spacecraft limited by single-use design, Starship's reusability and ability to refuel in space make it feasible to undertake repeated missions, transporting not only astronauts but also the vast quantities of supplies, infrastructure, and equipment necessary to establish long-term outposts.

For Mars, Starship opens the possibility of human settlement on an unprecedented scale.

With its ability to transport over 100 tons per mission, Starship can carry essential resources—such as habitats, life support systems, and energy sources—that are foundational to creating a self-sustaining colony. By drastically reducing launch costs through reusability, Starship enables more frequent missions, making it feasible to establish a continuous supply chain between Earth and Mars. This would allow settlers to receive regular deliveries of vital resources while developing local infrastructure, supporting SpaceX's vision of a city on Mars within the coming decades.

Similarly, Starship's capacity is ideally suited for lunar exploration. As NASA and other international agencies set their sights on building a lunar base, Starship can serve as a reliable means of delivering construction materials, scientific equipment, and even crew rotations to the Moon. Its reusability would enable a continuous cycle of lunar missions, each one building on the last, laying the groundwork for a permanent human presence. The Moon could become a hub for scientific research, technology testing, and even as a potential launch point for deeper space missions, all supported by regular Starship flights.

Beyond Mars and the Moon, Starship's capabilities extend into the realm of deep space. Equipped with in-space refueling, it could carry heavy scientific instruments or even entire orbital habitats to destinations in the outer solar system, including missions to the moons of Jupiter or Saturn. This capacity opens doors for in-depth exploration of worlds that were once considered unreachable, helping humanity gain insight into the solar system's most distant and mysterious regions.

The recent test flight of Starship isn't merely a technical milestone; it's a step toward a visionary future in which interplanetary travel becomes accessible and routine. Each successful flight, catch, and reusability improvement brings us closer to a reality where space travel is a regular part of human life. The implications are monumental. With reusable systems like Starship, humanity is not limited by the high costs and logistical restrictions that have historically defined space travel. Instead, space exploration can become a sustainable, scalable endeavor, supporting scientific breakthroughs, commercial ventures, and the expansion of human life beyond Earth.

The impact of Starship's development reaches beyond any single mission. It embodies a bold vision that pushes the boundaries of what we once thought possible, reshaping our ambitions for space. The road to becoming a multi-planetary species is no longer just a distant dream; with Starship, it is a tangible objective. Each test flight that refines its capabilities, each booster that returns safely, and each mission that advances reusability strengthens the foundation of a new era in space exploration. Starship's potential impact is not just in its destinations; it's in the realization of humanity's enduring desire to explore, survive, and thrive among the stars, turning the once unthinkable into a new normal.

Conclusion

The fifth flight test of Starship represents more than a technical achievement; it's a landmark in the journey toward making space exploration sustainable, scalable, and accessible. This mission has demonstrated that each element—from booster recovery and reusability to advanced heat shielding and landing precision—brings SpaceX closer to realizing the vision of affordable, routine space travel. Through each challenge overcome, each innovation achieved, and each lesson learned, SpaceX continues to prove that the dream of exploring and inhabiting other worlds is not only possible but within reach.

Reusability, a core principle of Starship's design, has the potential to revolutionize space travel by significantly lowering the costs associated with reaching orbit, the Moon, Mars, and beyond. Each successful test and recovery of the Starship system brings us closer to a future where spaceflight is no longer limited to singular, costly missions but becomes an iterative, economically viable endeavor. This reduction in cost could democratize space access, enabling scientists, industries, and explorers around the world to contribute to a new era of discovery and advancement. Space travel would no longer be confined to government agencies or isolated missions; it could become a shared human experience, shaping our collective future among the stars.

For those following SpaceX's journey, this flight test is an invitation to keep watching as boundaries are broken and history is written in real time. With each step forward, SpaceX redefines what humanity can achieve through persistence and innovation, showing that even the loftiest goals can be reached with the right combination of vision and resilience. The future of space exploration is unfolding before our eyes, and SpaceX's progress serves as a reminder of the endless possibilities when we dare to push beyond the familiar.

The story of Starship's development and success is a testament to the power of dreaming big. It reflects the spirit of exploration that has driven humanity to climb mountains, cross oceans, and now, reach for other planets. This journey is fueled by ingenuity and persistence, by the dedication of those who believe that the seemingly impossible can be achieved. As we look to the stars, Starship symbolizes not just the technology to reach them but the unwavering determination to go farther, discover more, and build a future that extends beyond the confines of our own planet. In this pursuit, SpaceX is showing the world that our capacity to innovate is boundless, and that with every challenge we overcome, we bring our dreams of interplanetary life one step closer to reality.

www.ingramcontent.com/pod-product-compliance
Lightning Source LLC
Chambersburg PA
CBHW070423240526
45472CB00020B/1164